Earth Changes

Dona Herweck Rice

Publishing Credits

Rachelle Cracchiolo, M.S.Ed., *Publisher*
Conni Medina, M.A.Ed., *Managing Editor*
Nika Fabienke, Ed.D., *Content Director*
Véronique Bos, *Creative Director*
Shaun N. Bernadou, *Art Director*
Carol Huey-Gatewood, M.A.Ed., *Editor*
Valerie Morales, *Associate Editor*
Courtney Roberson, *Senior Graphic Designer*

Image Credits: All images from iStock and/or Shutterstock.

Teacher Created Materials
5301 Oceanus Drive
Huntington Beach, CA 92649-1030
www.tcmpub.com

ISBN 978-1-4938-9818-3

It was  .

day

It is  .

night

It was cold.

It is  .

warm

It was  .

dry

It is  .

wet

It was  .

green

It is  .

brown

It was  .

young

It is  .

old

High-Frequency Words

is

it

was